HAVE YOU EVER SEEN
A SHELL WALKING?

HAVE YOU
EVER SEEN A
SHELL WALKING?

Odd Habitats of Aquatic Animals

Sarah R. Riedman

DAVID McKAY COMPANY, INC.
NEW YORK

Copyright © 1978 by Sarah R. Riedman

Library of Congress Cataloging in Publication Data

Riedman, Sarah Regal, 1902-
Have you ever seen a shell walking?

SUMMARY: Describes some of the more unusual habitats
where animals live, breed, find food, and receive
protection. Includes shells, sponges, and banks of a pond.
1. Animals, Habitations of—Juvenile literature.
[1. Animals—Habitations] I. Title.
QL756.R5   591.5′6   78-526
ISBN 0-679-20675-2

10 9 8 7 6 5 4 3 2 1

Manufactured in the United States of America

Illustration Credits

Department of Commerce, Florida News Bureau, page 30; Division of Photography, Field Museum of Natural History, Chicago, pages 38, 39; Division of Polar Programs, Polar Information Service, National Science Foundation, page 51; Environmental Studies Center, Jensen Beach, Florida, page 15; Karl Holland, Department of Commerce, Florida News Bureau, page 28; Michael W. Leavitt, Division of Polar Programs, Polar Information Service, National Science Foundation, page 44; Robert McGlynn, line drawings, pages 9, 11, 24, 26, 27, 32 bottom, 41; Marineland of Florida, pages 32 top, 33-35; Miami Seaquarium, pages x., 2-4, 7, 14, 20; National Oceanic and Atmospheric Administration, page 31; F. O'Leary, Division of Polar Programs, Polar Information Service, National Science Foundation, page 50; Eileen and Lou Rozee, Spongeorama, Tarpon Springs, Florida, page 22; Seashore Animals of the Pacific Coast, M.E. Johnson and H.J. Snook, reprinted through permission by Dover Publications, New York, pages 6, 17, 18; Official U.S. Navy photograph, by Photographer's Mate Airman (PHAN) Mark W. Huntley, USN, page 48.

Contents

PREFACE

MOST ANIMALS LIVE on land. Many live in the ocean, and these can live only in saltwater. Others make their home in fresh water—in a river, a lake, or a pond. Still others dig homes and live underground. And some—without a home of their own—are tenants inside another's dwelling.

The special home of each species of animal is its habitat, the place that suits its living needs. Some habitats may seem strange to us, but those habitats are where the animals find the kind of food they can eat, and where they can bear their young. They are places in which the animals and their babies are usually safe from natural enemies, and are protected from overly cold, hot, or dry conditions.

So read along about some habitats which may seem odd to us, but are just right for their inhabitants.

HAVE YOU EVER SEEN
A SHELL WALKING?

A hermit crab with its legs withdrawn.

Chapter 1

Home Is
a Borrowed Shell

HAVE YOU EVER seen a shell walking? You may see one sometime on the beach, or even find one on a concrete road near the ocean. If you look at one closely, you will see a pair of walking legs poking out of the shell's open end. They are the legs of a crab. The crab uses two pairs of walking legs as it travels sideways.

Two claws, one larger than the other, also stick out of the shell. The crab uses its claws for grasping food and carrying it to its mouth. If you pick up the shell, the crab will draw its legs inside it. Only the tips of the pincer claws will show, as if the crab is guarding the opening. This crab is one of a group called hermit crabs, or land crabs, because they search for food on land.

Like other crabs, the hermit starts its life in the

A hermit crab walks with its legs poking out of its shell.

ocean, where it lays its eggs. When they hatch into larvae, the larvae swim, then change and grow into infant crabs. At this stage, the crab has a hard cover on its back, like a tiny lobster. But it soon sheds its cover. This is called *molting.* After that happens, its abdomen curves under. Then the crab must find the empty shell of a dead snail, and crawl into it in order to protect its soft body. Hermit crabs of all sizes find shells, each to fit its own size. Small crabs crawl into periwinkle or moon snail shells. Larger ones use turban or whelk shells. The empty shells become their homes, and make it possible for them to wander safely onto land in search of food. But the shells are only temporary homes. Since the shells do not grow, the crabs must hunt for larger shells as their bodies grow. This usually happens when they are ready to start new generations of crabs. They must always protect their abdomens, to which their eggs attach.

Unlike its cousins, the hermit crab has almost freed itself from life at sea. It breathes air, and it is

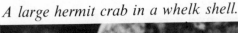
A large hermit crab in a whelk shell.

Note the eyes of this starry-eyed hermit crab.

not fussy about the food it finds on land, eating plant food or bits of flesh from dead or living animals. But it never wanders too far from seawater, where it releases its eggs or makes a quick change into a larger shell. While searching for food, the hermit crab is a landlubber, but it must go to sea to have its young.

Chapter 2

Life inside a Tube

ALONG THE SHORE above the tideline, many thousands of tiny animals find shelter in the rockweeds clinging to rocks. Some animals swim freely in tidal or rock pools. Others attach themselves to rocks. There you may find à small, snaillike creature—no more than one-half-inch wide—with a crown of fine threads, or tentacles, sticking out from the open end of the shell's coil. This animal is not a snail, but a plumed worm called *Spirorbis* (spy-roar'-biss). Its shell is really a coiled tube, built by the worm inside it. The worm settles in its home for life.

We usually think of a worm as being a long, slender, soft-bodied animal. But Spirorbis more closely resembles a snail. Another kind of tube-dwelling worm resembles a feather duster.

Tube-dwelling worms belong to a group known

Spirorbis lives in a coiled tube.

as annelids (an'-el-lidz), a name they got from the French word, *anneles,* meaning ringed ones. This is because the bodies of these worms are made of many rings or segments, which is why they are also called segmented or true worms. The clamworm and lugworm used for fish bait, the ragworm fed to aquarium fish, the earthworm, leeches, and sand-worms are other annelids.

The soft bodies of many kinds of worms make good eating for other animals, and so having hard tubes as homes means that they are safe from enemies.

Inside its tube, the Spirorbis breathes, takes in food and digests it, gets rid of wastes, and raises its young. The Spirorbis is both a chemist and an architect. It makes its hollow, spiral tube—its permanent home—from limelike chemicals secreted by cells in its skin. It then cements its home to a rock or

A feather duster tubeworm.

a clump of rockweed, so that the tube cannot be washed away by the incoming tide.

The worm's tentacles—those delicate, filmy threads on its head—are its life lines to the surrounding world. They serve as gills for breathing, by extracting oxygen dissolved in seawater. They also entangle bits of food floating nearby. The worm senses danger through its tentacles, and in time of danger it draws in its head. A neatly fitting lid, called the *operculum,* closes the tube's opening. Food is digested as it passes along a canal inside the tube. The worm's wastes are thrown off in little balls known as *castings.*

When the time comes to start a family, the Spirorbis fertilizes a batch of eggs in a brood chamber in the lowest coil of its tube. The eggs are laid out like beads on a string. Each string of a dozen or more eggs is wrapped in a thin tissue envelope resembling cellophane. When the eggs develop into larvae, the envelopes burst, and the young are propelled into the ocean. They swim about briefly in the intertidal zone—the area between high and low tides. Whatever happens next to the infant worms takes place between the incoming *(flood)* tide and the outgoing *(ebb)* tide. This is because they could be lost in a deep rising tide, too far from where they

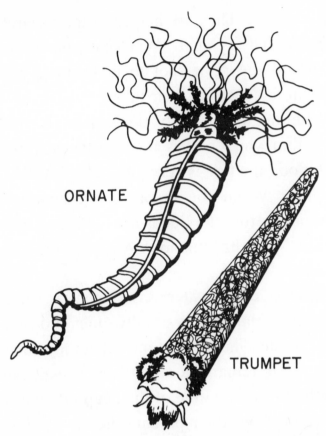

ORNATE

TRUMPET

Other species of tubeworms: 1. Ornate. 2. Trumpet.

must settle. During a receding tide, they run the risk of being washed up on shore where they cannot survive.

But the larvae swim about for only a short time. After an hour or so, they begin searching for a place in which to settle. It must be a spot close to where their parents released them, and where their ancestors also settled. They are helped in their search by their bright red eye spots. When they have instinctively found the right spot to make their homes, they begin to construct their tubes.

In addition to the daily tides, there is another tidal rhythm which occurs about twice each month. At full moon the tide is very high. This is called the *spring* tide. When the moon is one-quarter and three-quarters full, the tide is low. This is called the *neap* tide. What has this to do with the tube-building worm?

At neap tide the water's movement is sluggish, and the intertidal zone is usually quiet and safe for the tiny larvae. That is the time when they have the best chance to remain close to the rockweed. And it is exactly during the time of the neap tide that they are released from their parental home. During the same period, another batch of eggs is fertilized, and they begin developing into larvae which will be released at the next neap tide. By that time, the members of the previous generation are already safely settled in their own tube homes.

Another tube builder is the *Chaetopterus* (key-top'-tuh-russ), or parchment worm, that resides in a U-shaped tube. The tube is at least four times the length of its occupant, and is made from a parch-

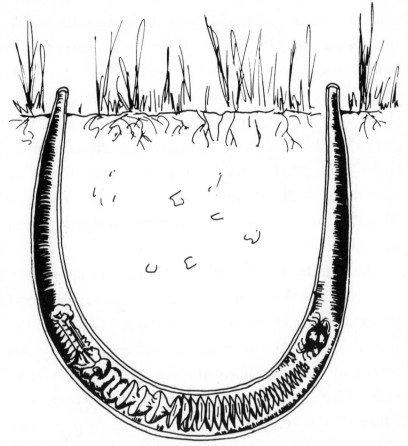

Chaetopterus, the parchment tubeworm, meets a pea crab—a non-paying tenant in the worm's home.

mentlike material secreted by the worm. The worm lives in the bend of the tube's U. The two arms of the U are open at the ends, and they stick up like chimneys above shallow water. These openings keep the worm in contact with the sea.

The Chaetopterus has three pairs of fan-shaped

11

paddles called *parapodia* (para-poh'-dee-ah), which push water past its body. This creates a current that brings in oxygen and food in the form of tiny floating plants. The plants are caught up in mucus, a thick slimy secretion. They are rolled into a ball ready for the next meal. Every twenty minutes or so, the Chaetopterus stops pumping water into one tube while it digests its food. The wastes then leave the tube through the other opening, and are carried away by the water's current.

The parchment worm has a non-paying tenant, the pea crab. While still young, the pea crab follows the steady stream of seawater carrying food into the worm's tube. As the pea crab grows, its body becomes too large to leave through the tube's narrow exits. The crab does no harm to the worm, and while it is receiving free shelter and food, it may also help to get rid of unwanted leftovers. Like a neighborly scavenger, the pea crab "sweeps" the tube clean for its host.

Chapter 3

Glued to a Rock

BARNACLES ARE ACTUALLY crustaceans, although they in no way resemble their crustacean relatives—lobsters, crayfish, crabs, and shrimp. Unlike their relatives, barnacles do not swim freely.

A barnacle is a stay-at-home, a prisoner inside a shell that is permanently anchored. The surf zone—below the surging waves that wash over the rocky coast—is home for many barnacles. Only on the ebb tide do the rocks, often coated with a colony of barnacles, come into view.

Barnacle shells are neither crushed by the pounding surf nor washed away by waves. They are firmly fixed to the rock by a cement their occupants make. Since a barnacle shell is shaped like a flattened cone, the waves roll over it.

Rocks are natural anchoring spots for barnacles,

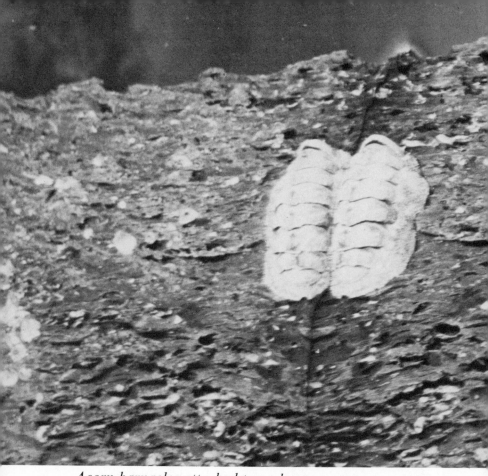

Acorn barnacles attached to rocks.

but they also attach themselves to other things, such as wharves, pilings, broken masts wedged between rocks, and floating timber. Sometimes barnacles even settle on a colony of mussels attached to rocks by the threads they spin. Some barnacles cling to ship bottoms, onto the hides of whales, or onto the shells of sea turtles for their entire adult lives.

The shell of a barnacle is made of five or six plates joined by small bits of shell. The shell's top is

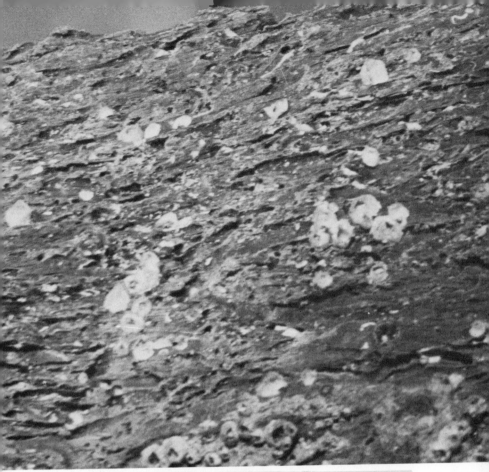

These acorn barnacles have settled on a colony of mussels which have attached themselves to rocks.

15

covered by two movable plates that form a lid. The barnacle can draw these plates apart in order to feed. At low tide, the barnacle closes the opening as if by two sliding doors. The closed plates protect the animal from drying out in the air. The plates are also shut by the occupant when it is alarmed by an approaching enemy.

Inside the shell, the barnacle lies on its back and sticks its feet, called *cirri* (seer'-ee), out of the opening. It washes food into its mouth by kicking its feet. It doesn't have to swim about in search of food, because the cirri rake in bits of floating plants and tiny animals, called *plankton*.

Barnacles come in different sizes, shapes, and colors. Their shells may range from a quarter of an inch to three inches in width. One species *(Balanus nubilus)*, living in Puget Sound, grows up to a foot in diameter. Some barnacle shells are almost colorless, while others are grayish, cream-colored, chalky white, or brilliant rose.

The two most familiar types of barnacles are the rock, or acorn, and the gooseneck barnacle. The gooseneck barnacle has a leathery stalk, or neck, which it attaches to a solid object. Its food-raking legs stick out of the opening at the top of its shell. It, too, has a movable lid for closing and opening.

Acorn barnacles.

It is only as adults that barnacles settle in one spot for life. As infants, the newly hatched barnacles swim about. But the babies in no way resemble the adults. When a baby barnacle leaves its egg, it is as free-moving as the larvae of other crustaceans. It has one eye, three pairs of legs, and a large mass of globules (or little globes) of fat. In this first larval stage, called *nauplius,* the barnacle doesn't have to search for food. It is nourished by the fat globules which also help to keep it afloat near the water's surface. When it is ready to change into its second

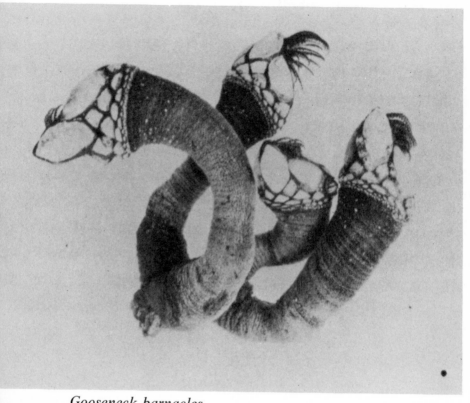

Gooseneck barnacles.

stage, called *cyprid* (sih′-prid), two more eyes, six more legs, a pair of head antennae, and a two-piece *(bivalve)* shell appear. It then becomes an adult.

The adult animal descends to the bottom of the sea floor in search of a suitable spot for a permanent home. It drags itself along until it finds a solid spot to which to attach itself by the first pair of head appendages. It prefers to make its home on a sur-

18

face that is a bit rough, rather than on a smooth surface or one covered with slimy seaweed. An ideal place might be where a colony of adults has already settled. Once the choice is made, the barnacle "sits down" on its head. There, the newly developed "glue pot," a cement-producing gland, releases a powerful adhesive, and the barnacle cements its head in place for life. It then begins to grow its stony shell, a small cup of lime enclosing its soft body.

Still other changes occur during the barnacle's life. Since it has nowhere to go, it loses its swimming legs. They are replaced by the branched and feathered cirri of adult life. Without gills, the barnacle breathes through these feathery hairs. It no longer has use for its eyes, and they are absorbed into the barnacle's tissues.

Then, safe in its home, the barnacle needs only to lie on its back and wait for the sea to bring it food. In a regular rhythm, its cirri sweep in currents of water that carry food toward its mouth. These sweeping movements of the cirri have been clocked by scientists. They beat with great regularity at about twenty times a minute.

Although hidden in its shell, the barnacle is not altogether safe from enemies. Fish, various kinds of

The bristle worm, a predator of barnacles.

bristle worms, and snails prey on barnacles. But the
worst predator of barnacles is the dog whelk, a snail
that drills a hole in the barnacle's shell or simply
pries open the shell's lid with its fleshy foot. It then
eats the soft-bodied barnacle. The empty barnacle
shell remains in the place to which it is attached,
and becomes a home for infant periwinkles, tube
worms, and young sea anemones. Or it may offer
temporary shelter to tidepool insects that have es-
caped from the rising tide.

One type of barnacle is a parasite that infests the
hides of whales. Another kind of barnacle attacks a

variety of crabs when the latter are still larvae and in the swimming cyprid stage. It pierces the crab's body and discharges a few cells into the crab's blood. The cells float in the crab's blood until they reach a spot near its intestine. There they attach themselves and grow, getting nourishment from the crab's blood. The cells then invade the crab's reproductive organs so that it can no longer produce baby crabs. Meanwhile, the barnacle has discharged its own eggs through the underside of the crab. The eggs develop into the nauplius stage and infect other crabs.

Barnacles can be a great nuisance to people when the animals encrust ship bottoms and increase the drag on ships in the water. This is why they are often called "fouling" organisms.

A sponge diver in full underwater gear boards a diving boat. He is holding a prong for pulling up sponges.

Chapter 4

An Apartment in a Sponge

A SPONGE DIVER uses a prong and underwater gear to pull up sponges from the sea floor. Sponges are animals that attach themselves for life to rocks, coral reefs, oysters, scallops, or abalone shells. They are the simplest of all the many-celled animals. They have no heads, tentacles, legs, mouths, gills, internal organs, blood systems, nerves, eyes, or ears.

A sponge has very little skeleton, but it supports clusters of cells and a system of canals and branching tunnels. This system, which reaches into every part of the sponge's body, is open to the sea. It connects with millions of tiny openings—pores in the sponge's slimy, leathery covering. Seawater is constantly sucked in through the pores and circulated through the canals. This canal system is the

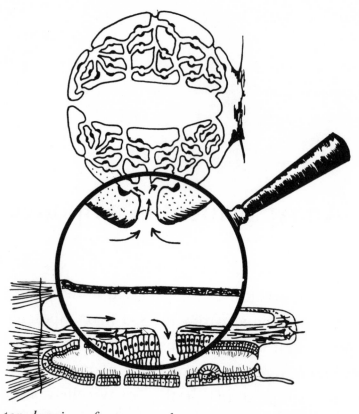

The top drawing of a sponge shows a system of canals, branching tunnels, and pores through which seawater is sucked in. The drawing at the bottom shows the inside of a canal containing lining cells with hairlike flagellae. Arrows show the direction of the current carrying food to the cells, and wastes out through the oscula.

reason why a sponge is able to hold water, whether it is used in the bath or to scrub walls or wash cars. But for a living sponge, the canal system is its means of staying alive. A sponge is made of masses of cells. It breathes, eats, grows, and produces eggs and sperm to create a new generation of sponges.

The cells that line the canals have small, hairlike extensions called *flagellae* (fla-jell'-eye), the Latin word for whips. The flagellae whip back and forth, creating a constant current for seawater that sweeps past the cells. The cells extract oxygen, and strain food—microscopic plants and animals—from the current. The wastes, sperm, and larvae are carried by the same current and flow out through a vent, called the *oscula*, at the top of the sponge. The water spurts out like a jet, as if the inside of the sponge was boiling over.

Sponges grow by adding on more cells. One kind of cell produces the supporting material, and some cells start a new generation of sponges. Both eggs and sperm are produced in the same sponge, but not always at the same time. The sperm cells are usually released into the water through the vent, and are later drawn in through the pores of another sponge. There the eggs are fertilized, and they develop into free-swimming larvae. When the larvae are released, they must find a suitable spot in which to settle—usually on a rock or an empty shell. Without a spot in which to settle, they die.

If the sponge settles on the shell of a live oyster or

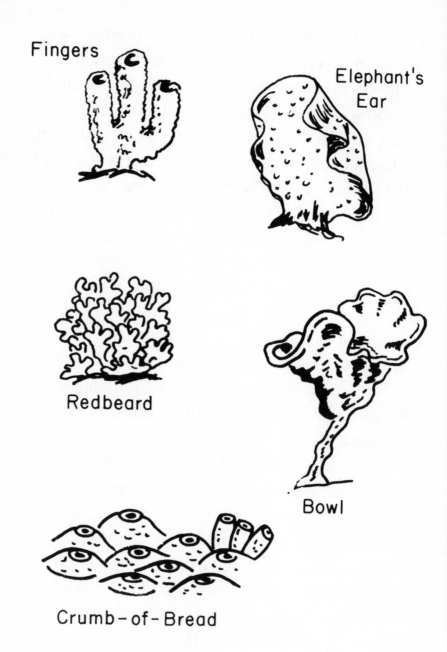

Fingers

Elephant's Ear

Redbeard

Bowl

Crumb-of-Bread

Sponges come in many shapes.

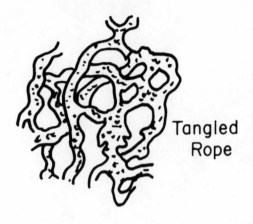

Tangled
Rope

Orange

Vase

Mitten

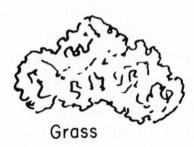

Grass

This boatman displays a giant loggerhead sponge at the Tarpon Springs, Florida, sponge docks.

clam, it kills the animal by boring into it. The sponge, however, can't eat the oyster or clam because it cannot digest flesh.

Once attached, the sponge grows into one of many different shapes, such as those resembling vases, bowls, pies, baskets, cakes, breadcrumbs, many-fingered gloves, trees, ropes, and elephants' ears. The adult sponges may be as small as a quarter of an inch, or as large as six feet in diameter.

28

The giant among sponges, known as the logger-head, attracts many small creatures which take up life in its canals and tunnels. Among these dwellers are shrimp, sea worms, hairy looking crabs, scallops, starfish, and even true fish. Some come to stay as permanent tenants that eat the food clinging to the walls of the sponge's passageways. Others are transients; they come and go, perhaps accidentally passing through.

The snapping, or pistol, shrimp is a permanent lodger. It measures about two inches long, and makes a sound by snapping the larger of its two claws. This popping sound can be heard coming from the inside of a sponge. In the open sea it is the shrimp's way of warding off an enemy. The adult shrimp remains inside a sponge tunnel, where it scrapes food bits from the sponge's walls, and bears its young. The baby shrimp, hatched from eggs, pass out of the sponge with the current. They swim or drift until they enter another sponge, where they find food and protection from enemies.

Other sponge occupants are shallow-water animals, too delicate to survive the wash of waves and inrushing tides. These animals are called *isopods* (eye'-so-podz) and *amphipods* (am'-fih-podz). Both are distant relatives of lobsters, crayfish, and crabs.

Visitors at the sponge docks in Tarpon Springs.

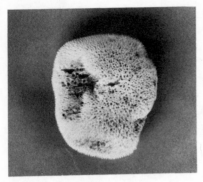
A bath sponge.

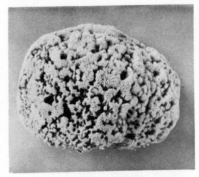
An anclote yellow sponge.

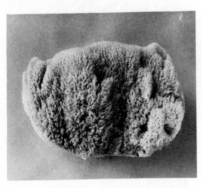
A Florida Key grass sponge.

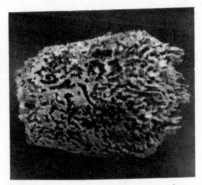
A Florida Key sheepswool sponge.

Isopods, which look like flattened bugs, are from a half inch to an inch long. They have crawling legs, a body with seven free-moving joints, and seven pairs of jointed legs. The female carries her eggs in a brood on the underside of her body. The hatched young, who look like their parents, do not go through a larval stage. Most isopods are scavengers that feed on dead sea animals. The food inside a

The snapping, or pistol, shrimp is a permanent sponge resident.

The hairy beach flea, an amphipod sponge resident.

sponge is likely to be suitable for isopods, and their apartment in a sponge is safe from small fish that would prey on them.

Amphipods are more shrimplike in shape. Some kinds swim and climb on underwater vegetation, from which they leap if the vegetation is disturbed. They may also hide in burrows of wet sand, out of reach of the small fish that would gobble them up. The sponge provides both safety and food for amphipods.

Fish also join the other sponge lodgers in what must be very crowded, but safe, quarters. Blennies and gobies, for instance, are bottom-dwelling fish, equipped with sharp teeth. They belong to a species

The goby, a sponge lodger.

called "cleaner fish," a name they earned because they make their living by actually cleaning and grooming such giant fishes as groupers, jacks, sea bass, and even moray eels. They pick clinging crustaceans from inside the fishes' mouths, and they nibble at parasites and peck away at infections that plague the larger fish. In return for these services, they receive their own meals.

What are blennies doing inside a sponge which

The sponge blenny lays and hatches its eggs in a sponge tunnel.

This sponge has completely overgrown the decorator crab.

does not require such services? Both male and female blennies give greater care to their eggs than do other fishes. While the male blenny guards the entrance to a sponge, the female lays the eggs inside it. The constant circulation of fresh seawater in a sponge's canals brings moisture and oxygen for the developing eggs. When hatched, the young swim out.

Sponge inhabitants don't harm their host. Sponges have few enemies because they have a repellent odor and taste. One kind of sponge shares its life with another animal, the hermit crab. The sponge surrounds the snail shell in which the crab makes its home. Sitting on the shell's "roof," the sponge is carried about to new feeding grounds. Both the crab and the sponge benefit by this arrangement. The crab is well camouflaged, and predators are kept away by its neighbor, the bad-tasting sponge.

Chapter 5

Holed Up in the Bank of a Pond

WE CALLED HIM "Max the Muskrat." He was a busy little fellow and a fast swimmer who would cross our pond in a few minutes. He swam directly toward the grassy bank on the opposite side of the pond, only to disappear underwater for a while. On his return trip, he always carried a clump of grass on his head. This he had pulled out of the pond's bank. Then he went under a heap of earth on the opposite bank. He made the journey so many times in one morning that it was hard to believe he had eaten so much grass. We wondered if he stored it away in his den pantry for another day.

Muskrats are well equipped for a life in water. Under their brown-red outer fur, they have warm and waterproof undercoats. Their hind feet, webbed like a duck's, are used as paddles. Their tails, almost

Muskrats are often seen on grassy banks of ponds.

The muskrat has a long, flat tail which he uses for steering in the water.

as long as their bodies, are bare of fur and flat. They serve as steering rudders.

An Indian legend describes the muskrat as an expert navigator who helped the Indian god *Nanabozho* (another name for Hiawatha) during the flood. For his services, the muskrat was allowed to choose where he would like to live. He selected the great blue lakes, but when he found that the lakes had no grasses on which to feed, he begged *Nanabozho* to change his home to fields and hills covered with grass. This *Nanabozho* did, but the muskrat still was not satisfied; he wanted to return once more to the lakes. The god became impatient, and said, "Muskrat, since you cannot make up your mind, your home will forever be in the marshes with plenty of grass to eat, and water deep enough for swimming."

This fable tells us something of the real life of muskrats. They live in fresh water ponds or salty marshes, and they build their houses with mud, reeds, and cattails. The finished product is a dome-shaped home with two rooms. The lower chamber, a kind of cellar, is under two feet of water. The upper room is in an earth mound two to four feet above the surface of the bank. Inside the room in the mound, the muskrat is hidden from its en-

The muskrat's two-storey underwater home.

emies—foxes, minks, otters, and fur trappers. The room is also used for sleeping and contains thick, warm beds of dry leaves and grasses. It is the place where the muskrat stores a supply of food, and it serves as a nursery for baby muskrats. A tunnel leads out of the lower chamber to permit the occupants to swim in and out of their home during winter when the pond has an ice cover.

Muskrats mate in winter and in early spring. A month after mating, the young, called *kits*, are born. There are usually four to twelve kits in a litter. The baby muskrats are blind and helpless for about a month. After that, they take care of themselves. When they are six months old, they are ready to

raise a family of their own, and they start to build a
home. In the banks of some marshes, there are
entire muskrat "towns," with "streets," dotted with
muskrat homes.

Chapter 6

At Home on the Frozen Continent

THE KANGAROO RAT makes its home in the hot desert. The emperor penguin's home is also in a desert, but one where the thermometer registers fifty degrees below zero. The desert is Antarctica, the coldest place in the world. It is a continent covered with never-melting ice, where snowfalls are rare. But it is just as dry as the parched, hot desert.

This frozen continent is at the bottom of the globe, so its seasons are reversed. The Antarctic summer lasts from December through February, and winter from late May through July. At the height of summer, there are no nights. In winter there is no daylight.

On this barren *permafrost* (permanently frozen ground), plants don't grow and animals have to go elsewhere for food. But the sea surrounding this

These emperor penguins have travelled inland on the Antarctic ice, about ten miles from the nearest land.

frigid land teems with living things of every kind, size, and shape. It is where the emperor penguin eats, grows fat, stores enough body fat for its stay on land, and finds food to feed its young.

The emperor penguin is one of seventeen species of penguins, all of which live in the southern hemisphere. Four species are found in the Antarctic. The others are distributed farther north, and one species is found as far north as the Galápagos Islands near the equator.

All penguins are sea birds that go onto land for only three reasons: to mate, to bear young, and to molt. They are more at home in the sea, where they feed and stay close to shore for about eight months of the year.

Fossils show that millions of years ago, penguin ancestors had wings and could fly. Later, penguins lost their ability to fly. Their wings are now short, paddle-like flippers, suitable for swimming. Their webbed feet act as rudders for steering. They keep warm in icy waters because they have a thick layer of fatty tissue under their feathers. And their feathers are water-proofed with oil.

The emperor penguin, the largest of all penguins, is a stately bird; the male stands four feet tall. Its short, thick legs support its weight, which is sometimes as much as eighty pounds. All penguins, but especially the emperor penguins, remind people of little fellows going to a party, dressed in formal evening clothes. That is how they appear in their upright positions, with their black feathers and dazzling white breasts and black-plumed hats. But the penguins' workaday outfits are worn the year round, except when they lose their feathers during molting periods.

Unlike the smaller species, which spend much of their time leaping about and romping on ice floes, the emperor penguins stand on the ice with their backs to the wind, huddled together against the cold. Expert divers and fast swimmers, they feed, rest while floating, and even sleep in the water. The tips of their feathers overlap, like the shingles of a roof. This arrangement forms a waterproof shell that protects the birds during their deep dives. Known to be the deepest divers among birds, they are able to dive deeper than people wearing scuba gear, and they can stay underwater for many minutes without breathing. Their breath-holding ability is explained by their unusual capacity for storing oxygen in their lungs, and for carrying greater amounts of oxygen attached to the hemoglobin (the red pigment in the blood) than most other birds.

Emperor penguins swim at nearly twenty miles an hour, almost as fast as the speediest sharks. They use up a great deal of energy by swimming most of the time. Even more energy is needed during the long trek on ice the birds take to reach the rookeries where they mate, breed and give birth.

The emperor penguin is well prepared for its aquatic life and the biting cold and blasts of wind on land. Not only is it snugly insulated with body fat and feathers, but researchers have found that the cells in its body burn body fuel at a very high rate. Scientists have measured the oxygen a penguin takes in and the carbon dioxide it exhales. The figures showed that its cell "furnaces," working full blast, furnish the large amount of energy the penguin needs for its strenuous life. The energy comes from the quantity of food the penguin eats.

The food of the emperor penguin is mainly fish and squid. Other penguin species in the Antarctic feed on krill, tiny shrimplike animals that live on the water's surface.

Emperor penguins also differ from the other Antarctic species in their breeding habits. They are the only birds that breed in winter. They build no nests, since there is just no nest-building material available. Each bird lays only one egg, which is deposited on solid ice. Late in March, the Antarctic autumn, the penguins arrive at their breeding ground. They return to the spot where they were born, sometimes waddling on the ice for fifty miles. For a change of pace, they flop on their bellies and

This Adélie penguin sits on her chick and an incubating egg at the same time in the midst of a crowded rookery. Unlike Emperor penguins, Adélie penguins hatch and raise their young during the relatively mild Antarctic summer.

slide as if they were toboggans. How they find the spot is a mystery, but for several seasons, tagged penguins have been known to return to the exact spot where they had been tagged as young chicks.

Finding a mate and courting sometimes takes until late in May or the middle of June. The males and females look so much alike that, when searching for a mate, a male penguin has trouble telling a male from a female. After mating, he will recognize her even if she is in the water some distance away. How? Another mystery.

As soon as the female has laid her single egg, the male starts incubating it. He places it on top of his feet, under the cover of a loose fold of belly skin which serves as a blanket. Because this fold of skin has a rich supply of blood-vessels, its temperature is higher than the rest of the penguin's body. He remains in an upright position and, with the egg on his feet, walks awkwardly to join the crowd of other penguins for protection against the cold. During this time, the female goes to sea in search of food. The male, however, does not eat, and may lose twenty pounds or more.

The chick hatches in about two months. By then (in August), the female has returned and relieves the male. Penguins take very good care of their

An Adélie penguin feeding a chick.

young. The babies, covered with fluffy coats of down, look like neglected little teddy bears with mussed-up brown fur. For several weeks they are helpless. They cannot swim and they need to be fed around the clock and to have their parents' warmth.

The parents digest the food and then bring it back up predigested. The babies suck the food out of the adults' mouths.

During their first winter, the emperor penguin chicks remain in the rookery under the care of a few adult babysitters, while their parents are off seeking food in the sea. When the young grow fat and become covered with feathers, the parents teach the

Molting emperor penguins.

chicks to swim so that they can take care of themselves.

Sometime during the Antarctic summer, the penguins molt—a process that takes several weeks. During the shedding of their beautiful feathers, they look very sloppy, and according to one naturalist, this condition seems to make them very sad. Deprived of much of their feathery covering, they cannot go to sea, and so they lose weight by not eating. Then, with freshly oiled new feathers, they return to sea, ready to feed, grow fat, and breed again.